Creatures of Legend

KRAKEN
GIGANTIC OCEAN TERROR

An Imprint of Pop!
popbooksonline.com

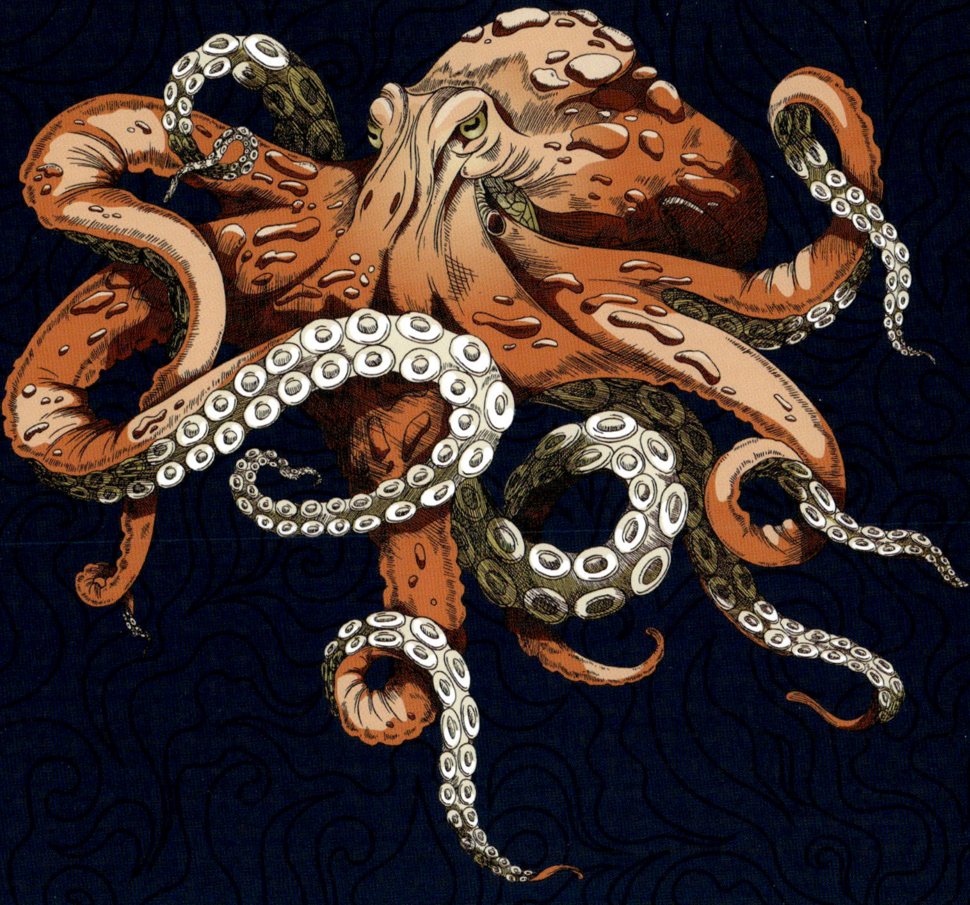

by Elizabeth Andrews

abdobooks.com

Published by Pop!, a division of ABDO, PO Box 398166, Minneapolis, Minnesota 55439. Copyright © 2023 by Abdo Consulting Group, Inc. International copyrights reserved in all countries. No part of this book may be reproduced in any form without written permission from the publisher. DiscoverRoo™ is a trademark and logo of Pop!.

Printed in the United States of America, North Mankato, Minnesota.

052022
092022

THIS BOOK CONTAINS RECYCLED MATERIALS

Cover Photo: Historia/Shutterstock

Interior Photos: Shutterstock Images, Stocktrek Images, Inc./Alamy Stock Photo, Science History Images/Alamy Stock Photo, Wikipedia Commons, Sarin Images/GRANGER, Alfred G. Mayer, Koji Sasahara/Shutterstock

Editor: Bridget O'Brien
Series Designer: Laura Graphenteen

Library of Congress Control Number: 2021951838

Publisher's Cataloging-in-Publication Data

Names: Andrews, Elizabeth, author.

Title: Kraken: gigantic ocean terror / by Elizabeth Andrews

Other title: gigantic ocean terror

Description: Minneapolis, Minnesota : Pop, 2023 | Series: Creatures of legend | Includes online resources and index

Identifiers: ISBN 9781098242350 (lib. bdg.) | ISBN 9781098243050 (ebook)

Subjects: LCSH: Kraken--Juvenile literature. | Giant squids (Monsters)--Juvenile literature. | Animals--Folklore--Juvenile literature. | Fabled creatures--Juvenile literature.

Classification: DDC 001.944--dc23

WELCOME TO DiscoverRoo!

Pop open this book and you'll find QR codes loaded with information, so you can learn even more!

Scan this code* and others like it while you read, or visit the website below to make this book pop!

popbooksonline.com/kraken

*Scanning QR codes requires a web-enabled smart device with a QR code reader app and a camera.

TABLE OF CONTENTS

CHAPTER 1
Beware the Kraken 4

CHAPTER 2
The Birth of the Kraken 10

CHAPTER 3
The Beast 16

CHAPTER 4
The Real Kraken 22

Making Connections 30
Glossary 31
Index 32
Online Resources 32

CHAPTER 1

BEWARE THE KRAKEN

The day started calm enough for the crew on a ship in the Norwegian Sea. It was the year 1575. The deck of the ship was busy as usual. A boy sat in the **crow's nest**, his eyes scanning the horizon for any dangers.

WATCH A VIDEO HERE!

Legends say that kraken are most active during the summer months.

A kraken's tentacles have spikes on them.

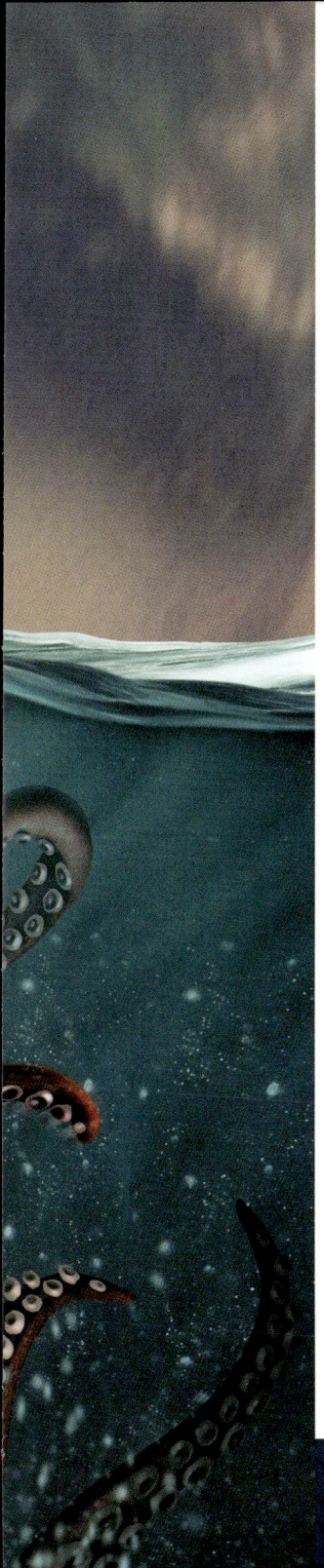

Suddenly, a section of water off the front of the ship was bubbling. Squinting at the movement, the boy saw thousands of fish swimming around in circles. The bubbles got bigger. The water grew darker. The boy knew what was happening before the first tentacle broke the surface. It was a kraken. The whole ship was doomed.

Hundreds of years ago, most boats were made of wood and were easy for kraken to destroy.

Within moments, the largest sea monster in the world was upon the ship. Its tentacles wrapped around the ship and snapped the wooden planks. The crew had nowhere to run. No one would survive.

Once the kraken had fed upon the crew, all that remained of the ship sunk to the sea floor. The kraken floated down too. It was time for it to rest again.

Many different kinds of sea creatures make their homes in sunken ships.

CHAPTER 2

THE BIRTH OF THE KRAKEN

Long ago, when science was not as advanced as it is today, legends and **zoology** merged. Research wasn't done the same way it is now. Things like submarines didn't exist. People couldn't get up close with underwater animals. When there was a

LEARN MORE HERE!

Not all sea monsters look the same.

quick sighting of something large and mysterious beneath the surface, people assumed it was a monster.

DID YOU KNOW? Sometimes when a kraken surfaced, people reported seeing whirlpools. These could easily sink boats!

Rumors of the kraken began after the fall of the ancient Greeks. However, the idea of the kraken might have been inspired by a Greek monster called the Scylla. She was 12 feet (3.7m) tall with six heads on **serpentine** necks. She lived in a deep-sea cave and devoured anything she could get her tentacles on.

THE FAMOUS KRAKEN

Alfred Lord Tennyson wrote a famous poem titled "The Kraken." It speaks about the slumbering ancient kraken at the bottom of the sea. The poem warns of the death the monster would cause when it finally rose to the surface. Tennyson used the kraken as a symbol of the unknown and how frightening and vast it could be.

Ancient Greeks believed in many kinds of monsters.

The first report of a kraken was made by King Sverre of Norway in 1180. Sverre spoke of a giant squid-like creature. It worried him. He warned sailors about all the monsters that guarded the shores of Norway. Sailors are considered **superstitious**. They feared the unknown, and the deep ocean was unexplored back then.

In the 1700s, a scientist named Carl Linnaeus recorded the kraken as an official animal. Linnaeus was an explorer

The spear-like tool used to catch large sea creatures is called a harpoon.

Scientists still organize plants and animals using the same style that Carl Linnaeus invented.

who studied nature. He wrote a book called *Systema Naturae* in 1735. The kraken was **classified** as a cephalopod. It has a soft body like an octopus or squid. Most sightings of the kraken were in the seas around Norway and Greenland.

THE BEAST

Many people traveled the Norwegian Sea in the 1700s. Most of the reports of the kraken were from traveling **clergymen** and sailors. Clergymen were some of the most educated people in science, math, and literature at the time. Sailors spent their whole lives learning

EXPLORE LINKS HERE!

about the sea. They would have the most firsthand knowledge of what lives within.

Kraken are often shown as giant octopuses.

No human who ever saw the entire body of a kraken lived to tell the tale.

There were many sightings of monstrous beasts out at sea until the 1900s. The descriptions ranged over the years. In 1734, Hans Egede reported a terrible creature coming up to his ship. It was bigger than the boat and had giant fins that helped it swim quickly through the water.

In a famous 1755 report, Bishop Erik Pontoppidan saw the "largest and most surprising of all animal creation." According to Pontoppidan the sea monster's whole body was too large for a human eye to see. He estimated it was a mile and a half wide. When he first laid eyes on the beast, he thought it was a small island. From the body, pointed horns rose out of the water as high as the **mast** of the ship. Thick tentacles floated around the body.

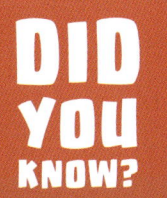

DID YOU KNOW? Many Marvel comic book superheroes have fought kraken in their stories.

Kraken could be as damaging to a ship as a storm could.

As time went on and more sightings occurred, people believed kraken were real creatures swimming in the sea.

However, they thought kraken were only 40 feet (12.2m) long, not a mile (1.6km).

The beasts spent most of their time on the sea floor. They would only come to the surface in July to **mate**. A kraken wouldn't break into the open air unless the water was perfectly calm.

The creatures always had large schools of fish following them. Smaller fish ate kraken waste. Sailors sometimes tried to catch a kraken, but they had never succeeded.

CHAPTER 4
THE REAL KRAKEN

With so many sightings of the kraken recorded through history, it is clear that there were giant creatures living in the dark, cold waters of the Norwegian Sea. As science advanced, mysteries about legendary creatures have cleared up. The kraken sightings were most likely giant squid.

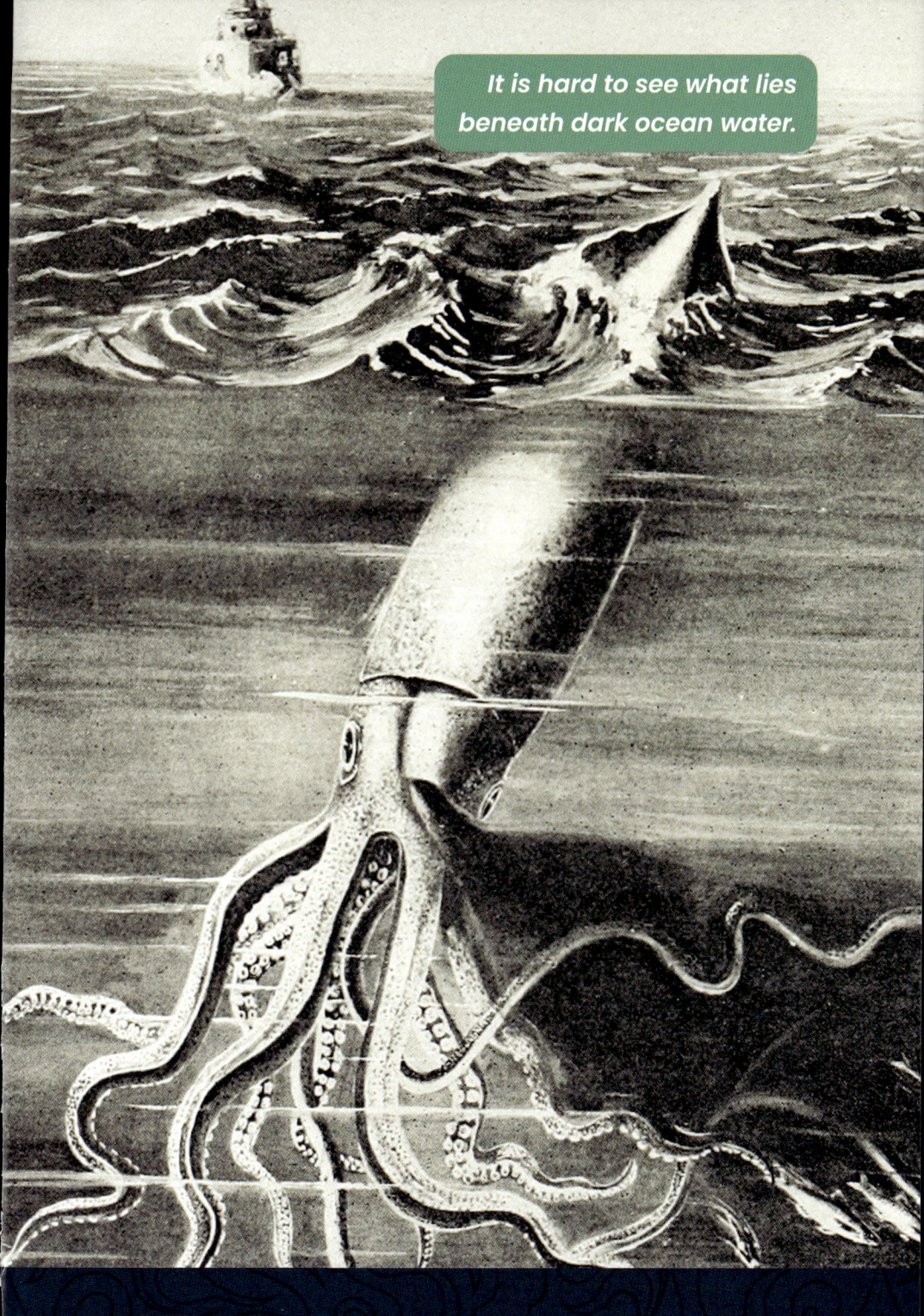
It is hard to see what lies beneath dark ocean water.

DID YOU KNOW? Giant squid eyes can grow to the size of basketballs. They must be large to see in complete darkness.

But even these beasts stay out of reach. Researchers have been struggling to find and study living giant squid. They live between 1,000 and 2,000 feet (305m and 610m) below the ocean's

The northern lights often shine above the Norwegian Sea.

surface. No sunlight reaches that deep, and the water is only 40°F (4.5°C). Exploring that area of the ocean is difficult. Most information about the giant comes from dead squid that washed ashore.

In 2006, this female giant squid was the first to ever be filmed alive.

The largest ever reported was 59 feet (18m) long and weighed 2,000 pounds (907kg). These real beasts have eight arms and two long feeding tentacles. The body and head of the squid, called the mantle, is about 6.5 feet (2m) long.

WHAT IS THIS GIANT?

Giant squid are massive. Their tentacles are covered with sharp-toothed suckers that can snatch prey from 33 feet (10m) away. These giant squid do not have many enemies in the ocean because they are so big. Whales have been discovered with scars from the sharp suckers of the beasts.

59 ft. long (18m)

25 ft. tall (7.6m)

4.5 ft. tall (1.4m)

Before people around the world had the scientific evidence to explain unbelievable sights, they formed answers for themselves. While the kraken did not turn out to be a monster from the pages

Kraken have been shown in movies and video games.

of a fantasy novel, the giant squid is fantastic by itself. Perhaps the difference between creatures of legend and a natural animal are not that far off.

MAKING CONNECTIONS

TEXT-TO-SELF

Do you think there are more unknown creatures living deep under the sea? If so, what do you think they look like?

TEXT-TO-TEXT

Have you read any stories that had sea monsters in them? If so, what did those monsters have in common with the kraken?

TEXT-TO-WORLD

Why do you think some people fear the ocean?

GLOSSARY

classified — assigned to a certain group.

clergyman — an official leader of the Christian church.

crow's nest — a platform high on a ship's mast used as a lookout.

mast — a tall, supportive pole that rises from the deck of a ship.

mate — to come together to have young.

serpentine — having twisting form or movement; snakelike.

superstitious — willing to believe things without true evidence.

zoology — the science and study of animals.

INDEX

giant squid, 22–27, 29

Linnaaus, Carl, 14–15

Norway, 13, 15

Norwegian Sea, 4, 15, 16, 22

science, 10, 16, 22, 28

Scylla, 12

sea floor, 9, 12, 21

sightings, 13, 15–16, 18–20, 22–23

Tennyson, Alfred Lord, 12

tentacles, 7–8, 12, 19, 26–27

ONLINE RESOURCES
popbooksonline.com

Scan this code* and others like it while you read, or visit the website below to make this book pop!

popbooksonline.com/kraken

*Scanning QR codes requires a web-enabled smart device with a QR code reader app and a camera.